Math

Written by

Sharon L. Apichella and Mary D. Sutton

Editor: Gillian Snoddy
Cover Illustrator: Gloria Jenkins
Designer/Production: Alicia Triche
Art Director: Moonhee Pak
Project Director: Stacey Faulkner

Table of Contents

Introduction

Cootie Catchers Math is an interactive and motivating tool for daily skill review. Using a new twist on the popular origami fortune tellers, this hands-on resource provides a fun and unique approach to practicing and reviewing standards-based math concepts and academic language. *Cootie Catchers Math* features 20 reproducible cootie catchers that each reinforce a specific math skill. Each page includes a *Before You Flip* hint for students to apply while they use each cootie catcher and an *After You Flip* activity to extend their learning after they have finished. Once the cootie catcher is made, students read and solve the problems; then they lift the flaps to reveal the correct answers. A recording sheet is provided on the inside back cover to help teachers keep track of assigned cootie catchers.

Aligned to the National Council of Teachers of Mathematics (NCTM) standards, *Cootie Catchers Math* is an ideal resource for providing specific content review for all students. Research shows that repetition is essential for the brain to learn and recall information. Furthermore, children have a tendency to repeat activities they enjoy. *Cootie Catchers Math* offers a fun and quick way for students to repeat and retain essential skills. This teacher-tested, student-approved resource can be used for classroom center activities, as enrichment assignments when regular class work is completed, or for homework. Perfect for individuals, partners, or small groups, *Cootie Catchers Math* makes practicing math skills enjoyable. The following areas are addressed in this resource:

- Place value
- Rounding
- Addition
- Subtraction
- Multiplication
- Division
- Fractions
- Decimals
- Algebra
- Geometry
- Time
- Money
- Measurement
- Problem solving

Cootie catchers fit in pants pockets, backpacks, or lunch boxes for review on the go! Students can use them in a classroom center, at their desks, on the playground, or in a car or bus. Parents can slip Cootie catchers into a pocket or purse and use them to review with their child at home, in line at the store, or while waiting for appointments. With these easy-to-make, fun-to-use, portable manipulatives, students will love reviewing math skills and vocabulary the *Cootie Catchers Math* way!

Getting Started

How to Use

1. Select a skill you would like your students to practice, and make multiple copies of the corresponding page. Store the pages in a labeled hanging file in a math center or where math manipulatives are stored.
2. Demonstrate how to fold the Cootie Catchers. Display the instructions for students' reference.
3. Remind students to read the *Before You Flip* section before using each cootie catcher.
4. Have the students complete the *After You Flip* activity as an extension or quick assessment after they have used each cootie catcher. Ask the students to return the top portion of the page to you. Use this, along with the recording sheet, to keep track of assigned cootie catchers.
5. Send the cootie catchers home for additional practice.

How to Make

1. Carefully cut along the outline of the square. Fold and unfold the square in half diagonally in both directions to make two creases that form an X.

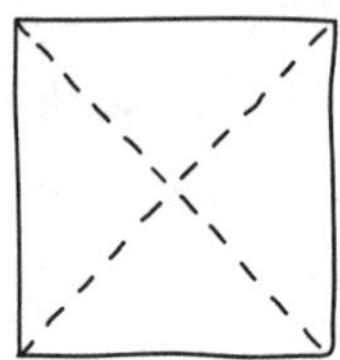

2. Place the paper facedown, and then fold each of the four corners in so that their points touch the center.

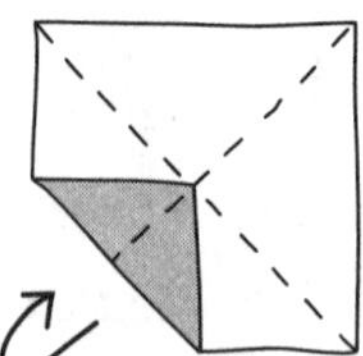

3. Turn the paper over so the flaps are facedown. Again, fold each of the four corners in so their points touch the center.

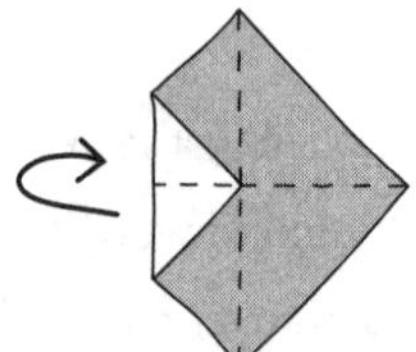

4. Fold the square in half, making a rectangle. Unfold and fold in half in the opposite direction, making a rectangle.

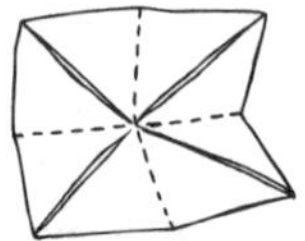

5. Slide both index fingers and thumbs under the four flaps.

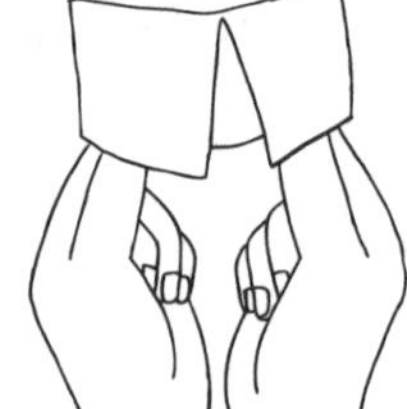

6. Use your thumbs and index fingers to pinch the top corners together and form a point. You are ready to play.

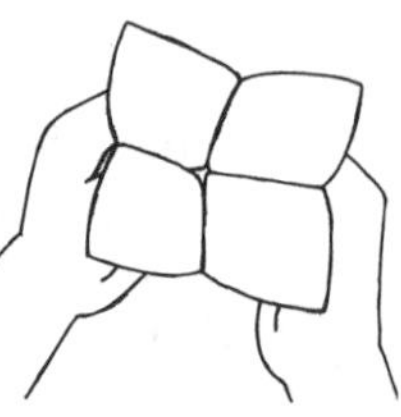

How to Play

1. Choose a number from one to five.
2. Open and close the cootie catcher (front to back and then sideways) as many times as the number selected.
3. Choose one of the four questions shown inside and answer it.
4. Lift the flap on which the question is written and check the answer.
5. Continue playing in the same way until all eight questions have been answered.

Name ______________________ Date ______________________

Value of the Underlined Digit

Before you "FLIP"

Millions				Thousands				Hundreds	Tens	Ones
hundred millions	ten millions	millions	,	hundred thousands	ten thousands	thousands	,	hundreds	tens	ones
8	3	6		2	1	0		9	4	5

After you "FLIP"

Write the following number in word form: 56,890,304

____

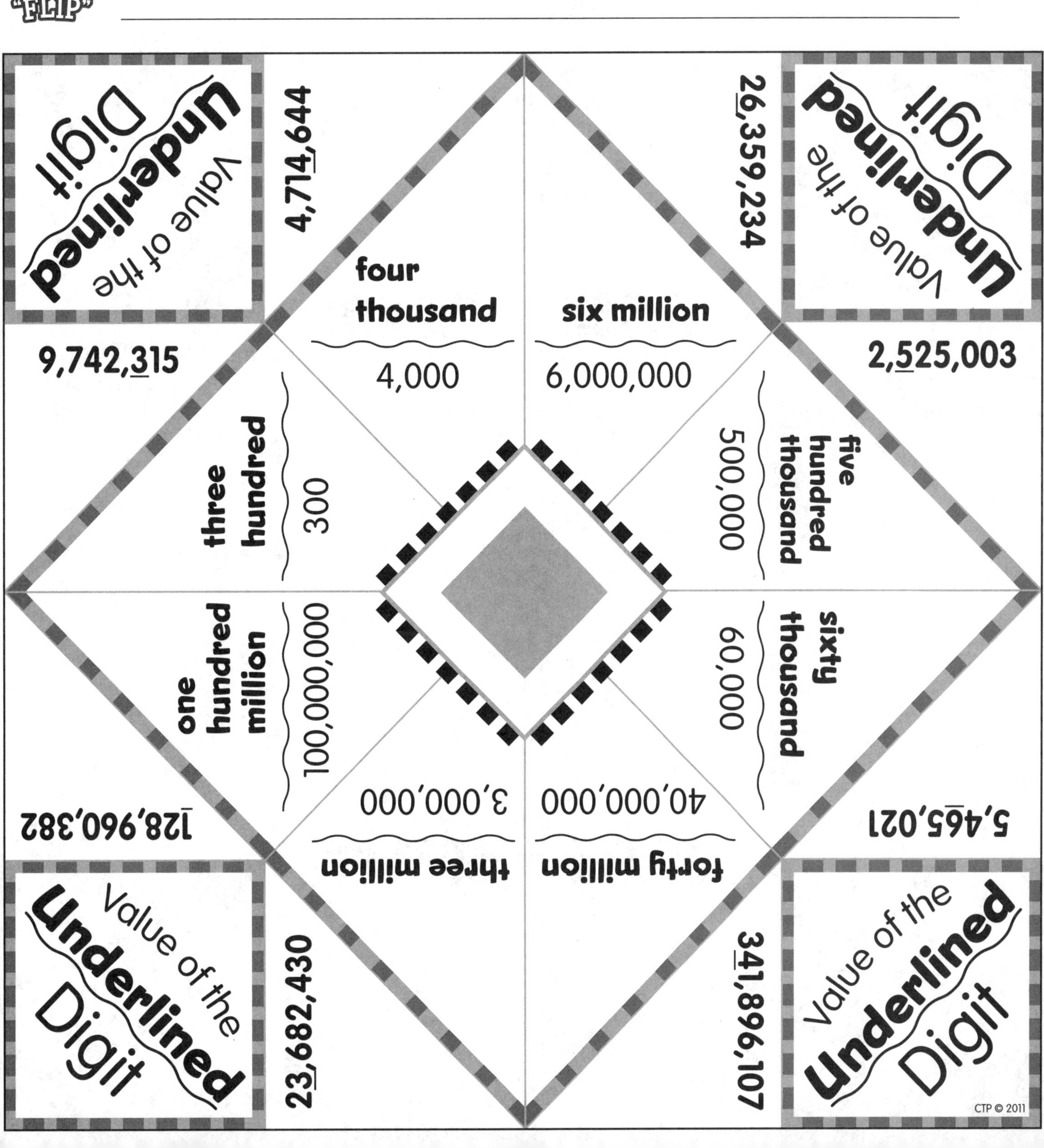

Name ______________________________ Date ______________

Rounding Numbers

Hint: Determine your rounding digit (the digit you need to round to). If the digit to the right of the rounding digit is less than 5, do not change the rounding digit. If it is 5 or greater, add one to the rounding digit. Round 6,342 to the nearest hundred. 6,③42 → 6,300

After you "FLIP"

Round 487,858,039 to the nearest ten thousand, hundred, and ten.

______________ ______________ ______________

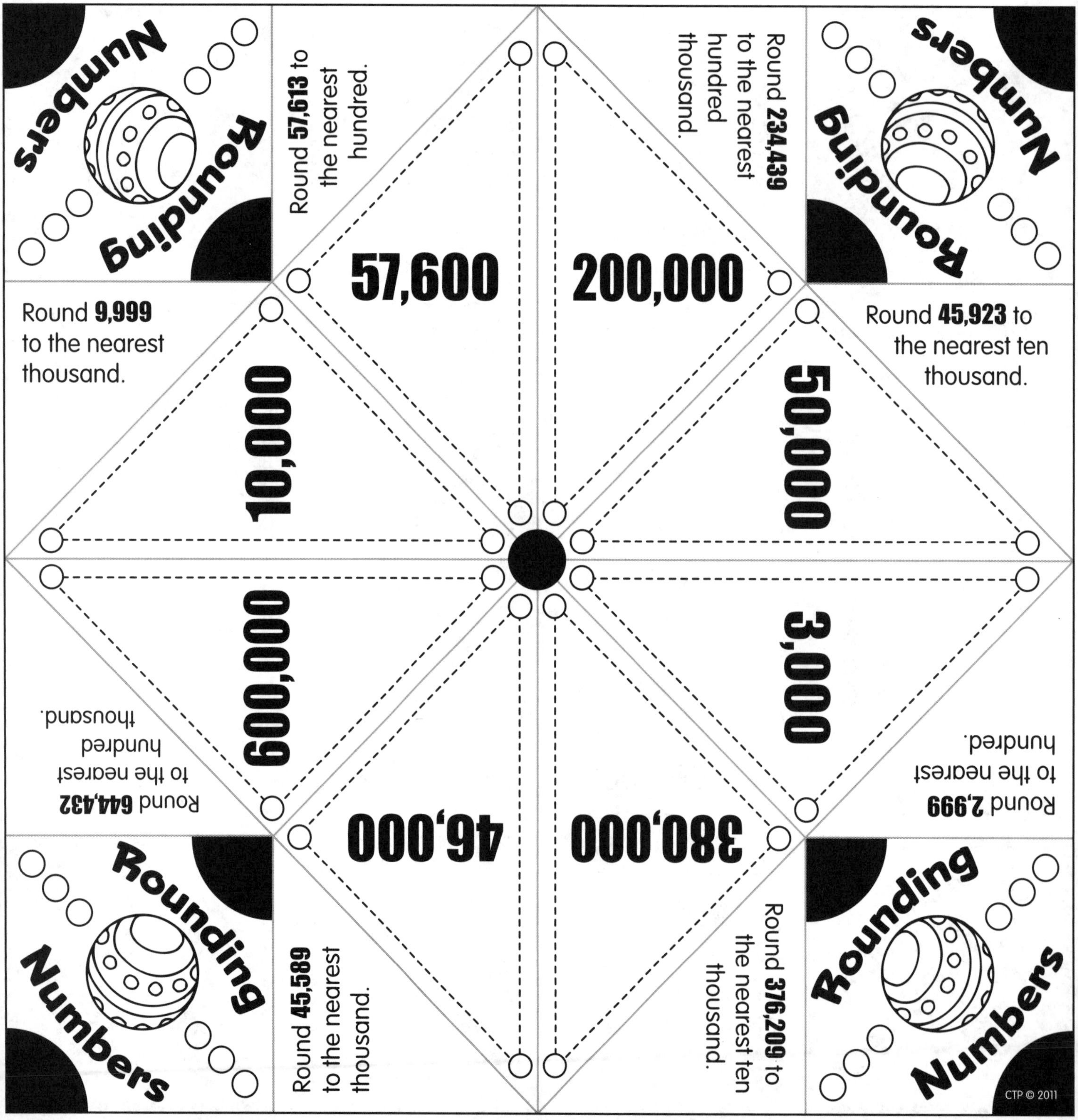

Rounding Numbers

Round **57,613** to the nearest hundred.

57,600

Round **234,439** to the nearest hundred thousand.

200,000

Rounding Numbers

Round **9,999** to the nearest thousand.

10,000

Round **45,923** to the nearest ten thousand.

50,000

Round **644,432** to the nearest hundred thousand.

600,000

Round **2,999** to the nearest hundred.

3,000

Rounding Numbers

Round **45,589** to the nearest thousand.

46,000

Round **376,209** to the nearest ten thousand.

380,000

Rounding Numbers

Name ____________________ Date ____________________

Mean, Mode, Median, and Range

Before you "FLIP"

Hint: Mean is the sum of the numbers divided by the number of addends. **Mode** is the number that occurs most often. **Median** is the middle number of a set that is in numerical order. **Range** is the difference between the greatest and least numbers.

After you "FLIP"

Find the mean, mode, median, and range for the following test scores: 75, 98, 90, 100, 87

Mean = ________ Mode = ________ Median = ________ Range = ________

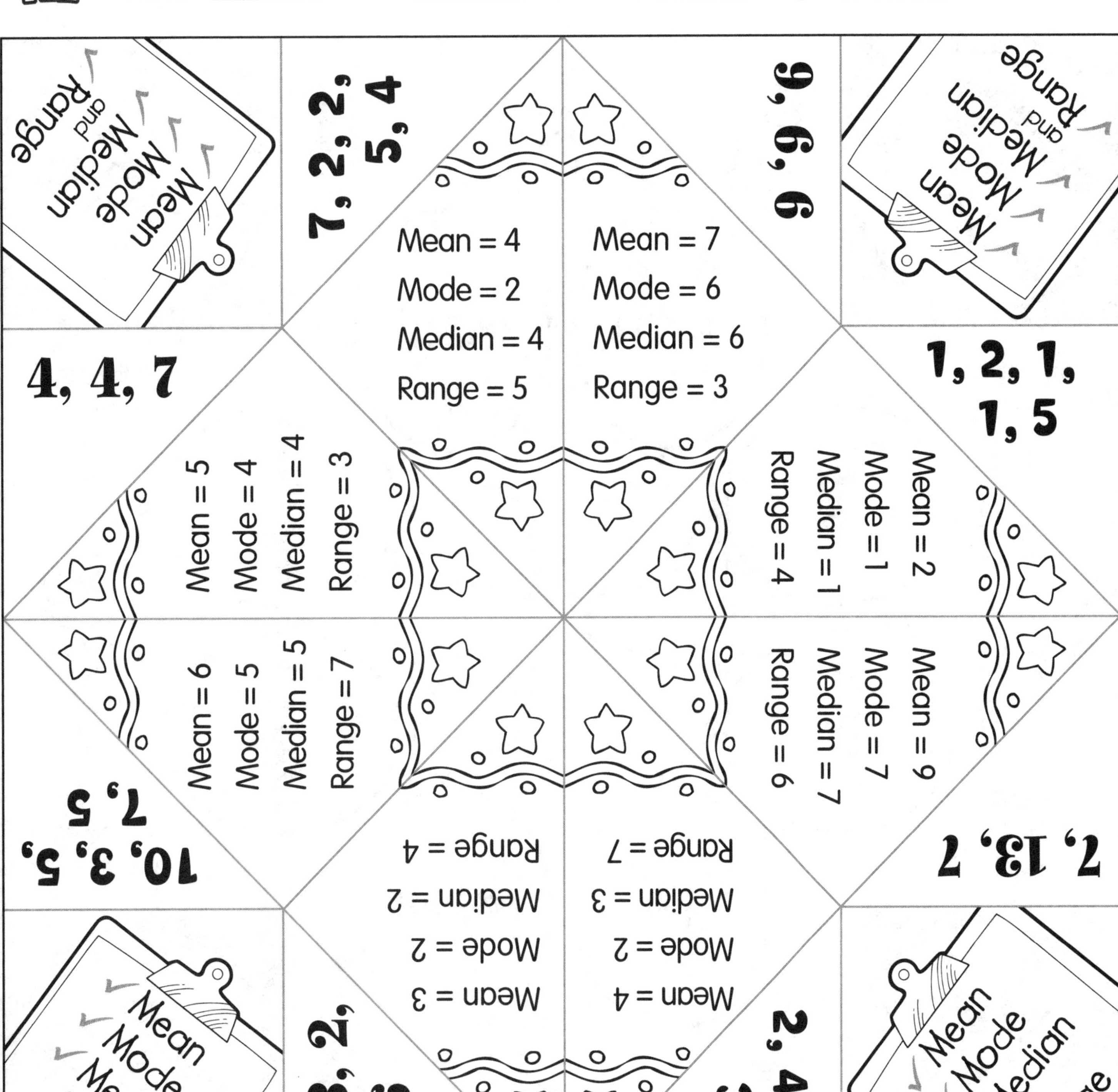

 Name ______________________ Date ______________

Fractions of Words

Hint: Count the number of letters in each word when solving these problems. Name the first $\frac{1}{5}$ of the word *solve*. The first of the five letters is *s*.

Write the fraction that represents the number of vowels in the word *arithmetic*.

____

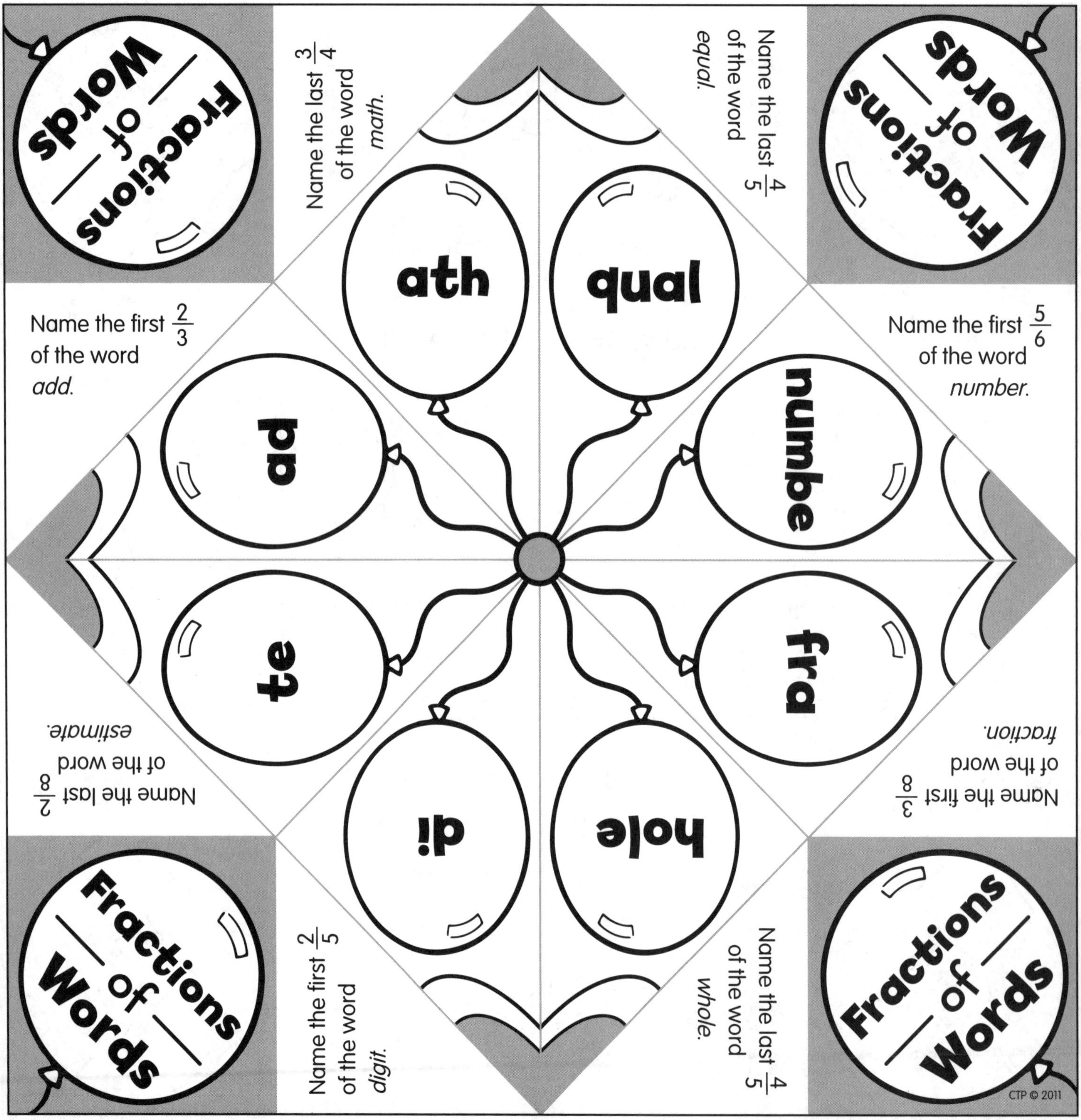

Fractions of Words

Fractions of Words

Name the last $\frac{3}{4}$ of the word *math*.

Name the last $\frac{4}{5}$ of the word *equal*.

Name the first $\frac{2}{3}$ of the word *add*.

Name the first $\frac{5}{6}$ of the word *number*.

ath

qual

ad

numbe

te

fra

di

hole

Name the last $\frac{2}{8}$ of the word *estimate*.

Name the first $\frac{3}{8}$ of the word *fraction*.

Name the first $\frac{2}{5}$ of the word *digit*.

Name the last $\frac{4}{5}$ of the word *whole*.

Fractions of Words

Fractions of Words

Name ______________________________ Date ______________________

Fractions in Simplest Form

Hint: To find the simplest form of a fraction, divide until 1 is the only number that divides both the numerator and the denominator.

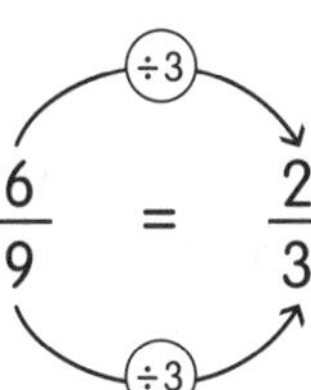

Is $\frac{9}{15}$ in simplest form? ___________

Explain. ____

$\frac{\square}{12} = \frac{1}{4}$

$\frac{4}{8} = \frac{\square}{2}$

Fractions in Simplest Form

$\frac{3}{12} = \frac{1}{4}$

$\frac{4}{8} = \frac{1}{2}$

$\frac{15}{20} = \frac{\square}{4}$

$\frac{\square}{18} = \frac{1}{3}$

$\frac{15}{20} = \frac{3}{4}$

$\frac{6}{18} = \frac{1}{3}$

$\frac{12}{14} = \frac{6}{7}$

$\frac{6}{15} = \frac{2}{5}$

$\frac{12}{14} = \frac{\square}{7}$

$\frac{6}{15} = \frac{2}{\square}$

$\frac{16}{20} = \frac{4}{5}$

$\frac{10}{14} = \frac{5}{7}$

Fractions in Simplest Form

$\frac{\square}{20} = \frac{4}{5}$

$\frac{10}{\square} = \frac{5}{7}$

Fractions in Simplest Form

Name ____________________ Date ____________________

Comparing Fractions

Hint: To help you compare fractions with different denominators, draw or visualize the fractions.

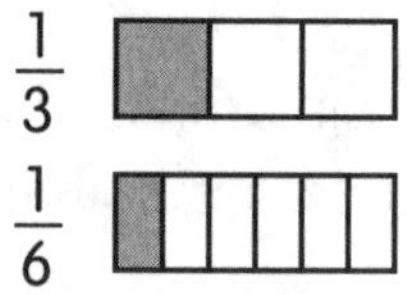

$\frac{1}{3}$

$\frac{1}{6}$

Is $\frac{2}{5}$ greater than $\frac{3}{4}$? __________

On the back of this paper, draw the fractions to support your answer.

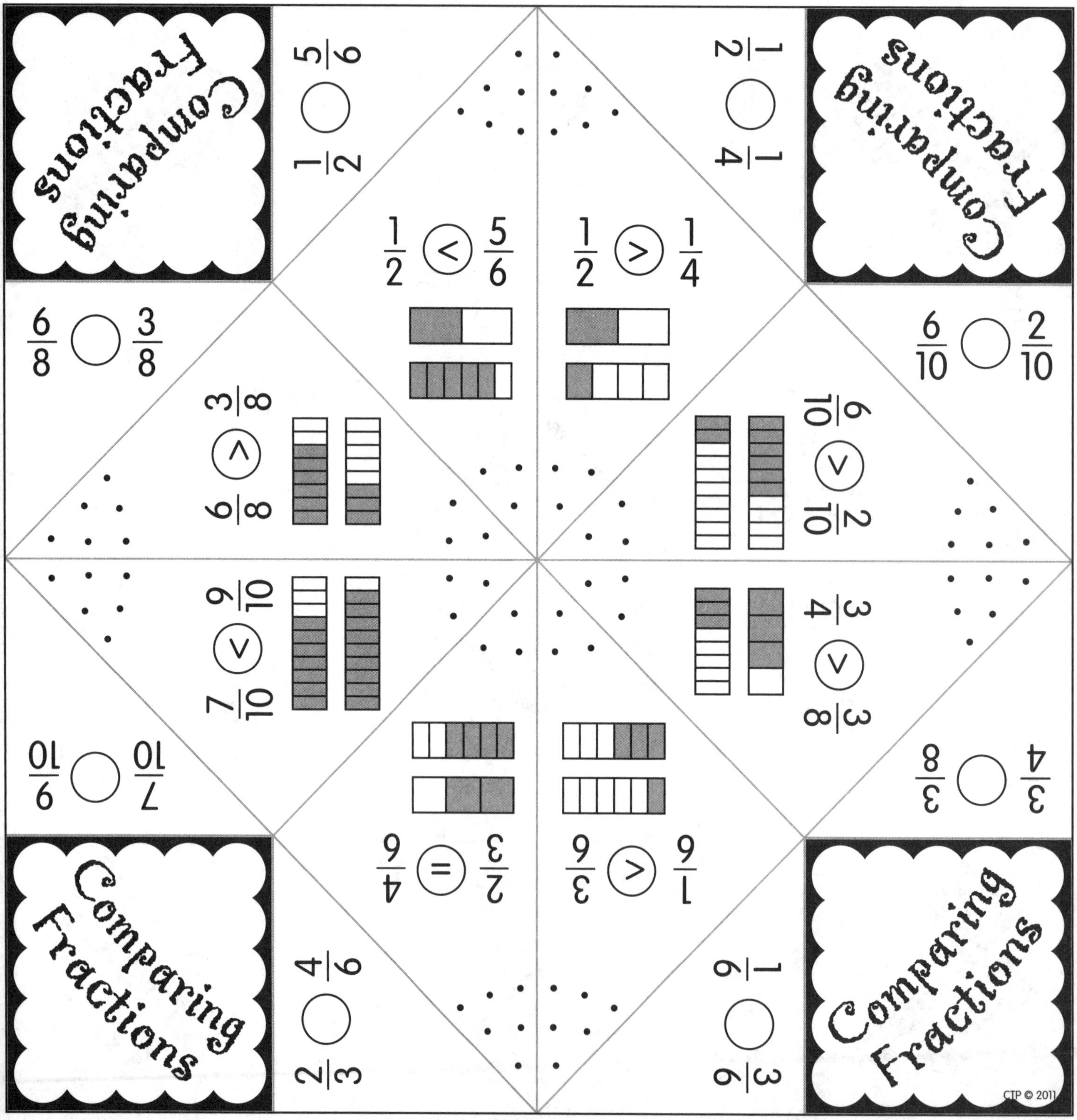

Comparing Fractions

$\frac{1}{2}$ ◯ $\frac{5}{6}$

$\frac{1}{2}$ ◯ $\frac{1}{4}$

Comparing Fractions

$\frac{1}{2}$ (<) $\frac{5}{6}$

$\frac{1}{2}$ (>) $\frac{1}{4}$

$\frac{6}{8}$ ◯ $\frac{3}{8}$

$\frac{6}{10}$ ◯ $\frac{2}{10}$

$\frac{6}{8}$ (>) $\frac{3}{8}$

$\frac{6}{10}$ (>) $\frac{2}{10}$

$\frac{7}{10}$ (<) $\frac{9}{10}$

$\frac{3}{4}$ (>) $\frac{3}{8}$

$\frac{7}{10}$ ◯ $\frac{9}{10}$

$\frac{3}{4}$ ◯ $\frac{3}{8}$

$\frac{2}{3}$ (=) $\frac{4}{6}$

$\frac{3}{6}$ (>) $\frac{1}{6}$

Comparing Fractions

$\frac{2}{3}$ ◯ $\frac{4}{6}$

$\frac{3}{6}$ ◯ $\frac{1}{6}$

Comparing Fractions

Name ______________________ Date ______________________

Fraction Word Problems

Before you "FLIP" **Hint:** A fraction describes part of a group or part of a whole.

After you "FLIP" What fraction represents the number of girls in your classroom? _________

Boys? _________

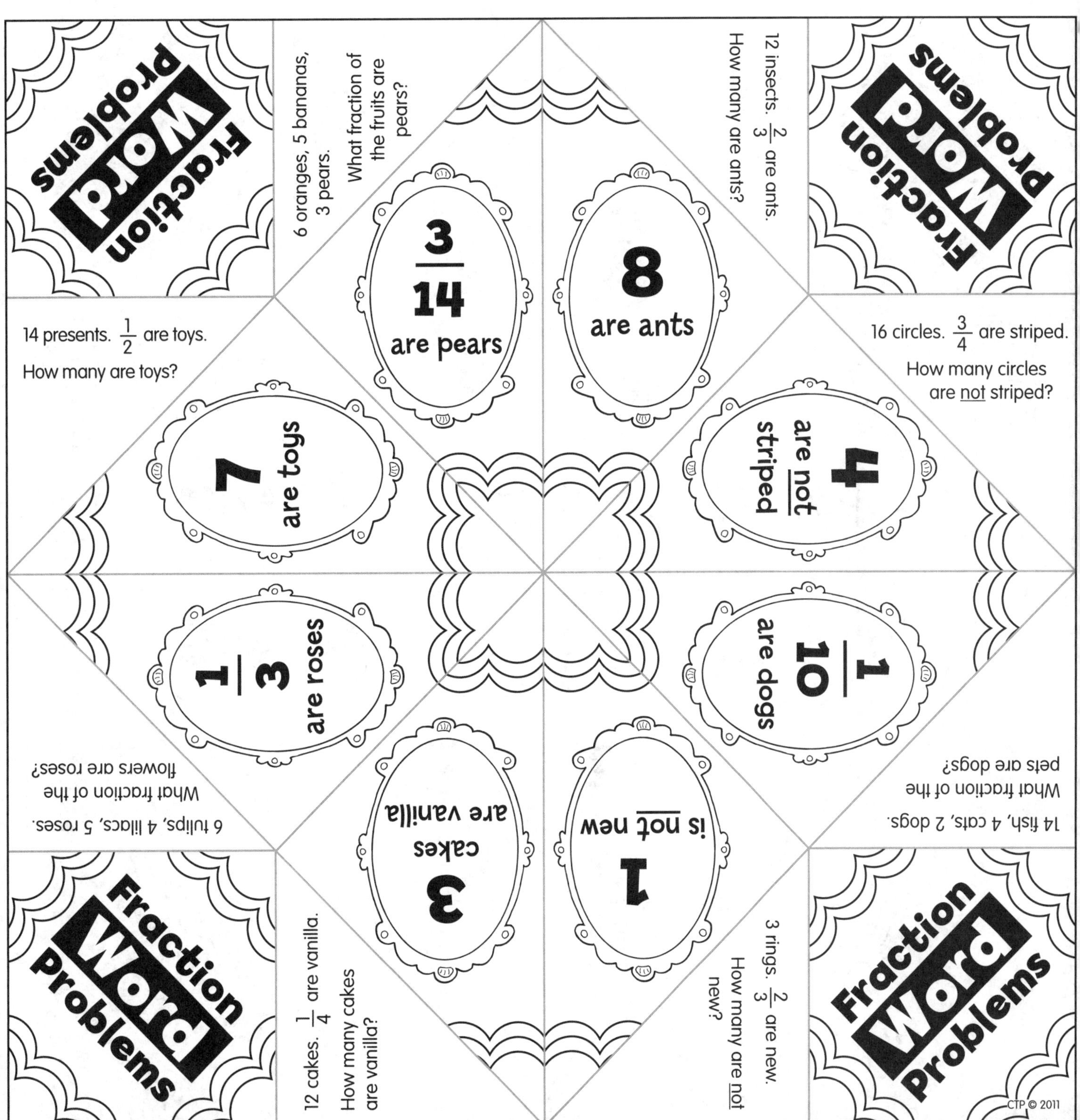

 Name ______________________ Date ______________

Improper Fractions and Mixed Numbers

Hint: An improper fraction is a fraction greater than or equal to 1. Its numerator is greater than or equal to the denominator. A mixed number is represented by a whole number and a fraction.

Improper fraction: $\frac{5}{4}$ Mixed number: $1\frac{1}{4}$

What fraction is equivalent to $2\frac{4}{5}$? ______________

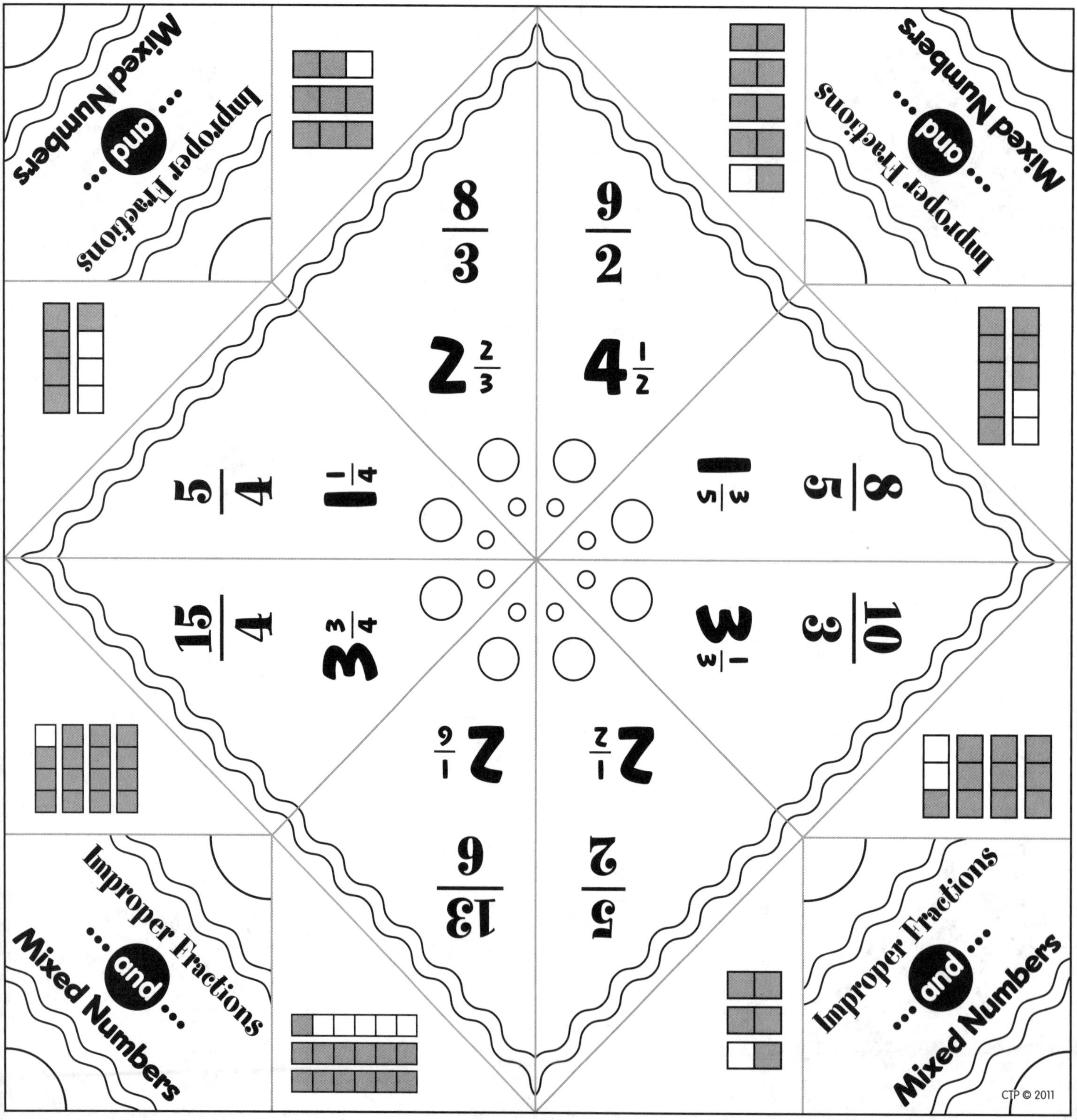

Improper Fractions ...and... Mixed Numbers

$\frac{8}{3}$ $2\frac{2}{3}$

$\frac{9}{2}$ $4\frac{1}{2}$

$\frac{5}{4}$ $1\frac{1}{4}$

$\frac{8}{5}$ $1\frac{3}{5}$

$\frac{15}{4}$ $3\frac{3}{4}$

$\frac{10}{3}$ $3\frac{1}{3}$

$\frac{13}{6}$ $2\frac{1}{6}$

$\frac{5}{2}$ $2\frac{1}{2}$

Improper Fractions ...and... Mixed Numbers

Improper Fractions ...and... Mixed Numbers

Improper Fractions ...and... Mixed Numbers

Name ____________________ Date ____________________

Order the Decimals

Hint: When ordering numbers with decimals, start at the left and find the first place where the digits differ. Then compare the decimals.

After you "FLIP"

Write the missing decimals: __________, 5.62, 5.63, __________, 5.65, __________

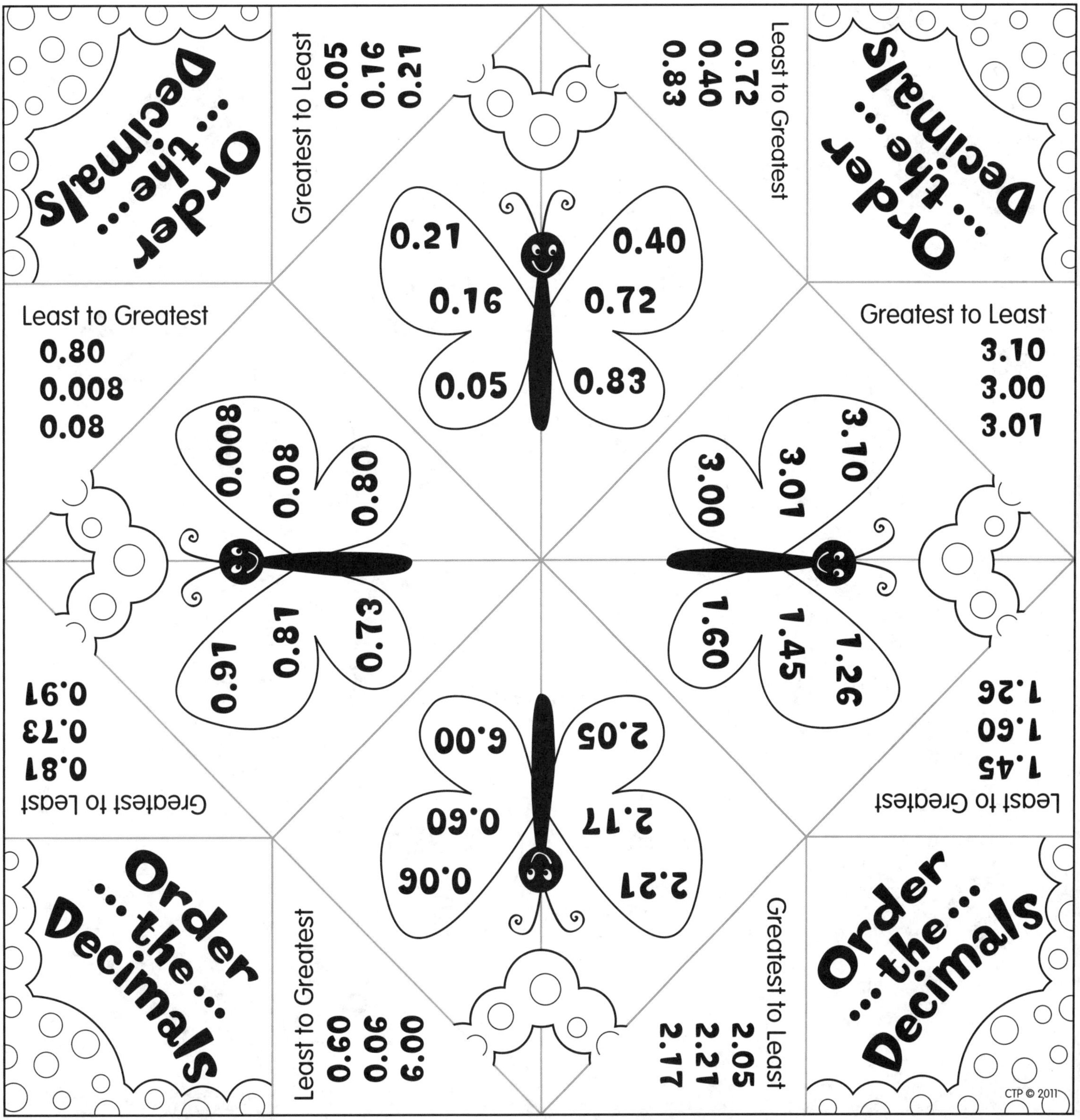

Order ...the... Decimals

Greatest to Least
0.05
0.16
0.21

Least to Greatest
0.72
0.40
0.83

Order ...the... Decimals

0.21
0.16
0.05

0.40
0.72
0.83

Least to Greatest
0.80
0.008
0.08

Greatest to Least
3.10
3.00
3.01

0.008
0.08
0.80

3.10
3.01
3.00

0.91
0.81
0.73

1.26
1.45
1.60

Greatest to Least
0.81
0.73
0.91

Least to Greatest
1.45
1.60
1.26

6.00
0.60
0.06

2.05
2.17
2.21

Order ...the... Decimals

Least to Greatest
0.60
0.06
6.00

Greatest to Least
2.05
2.21
2.17

Order ...the... Decimals

Name ______________________________ Date ____________________

What's the Rule?

Before you "FLIP" **Hint:** If the value of the numbers in the series is increasing, you are adding or multiplying. If the value is decreasing, you are subtracting or dividing.

After you "FLIP" Write the rule and the missing number for the following: 200, 100, 50, ________

____

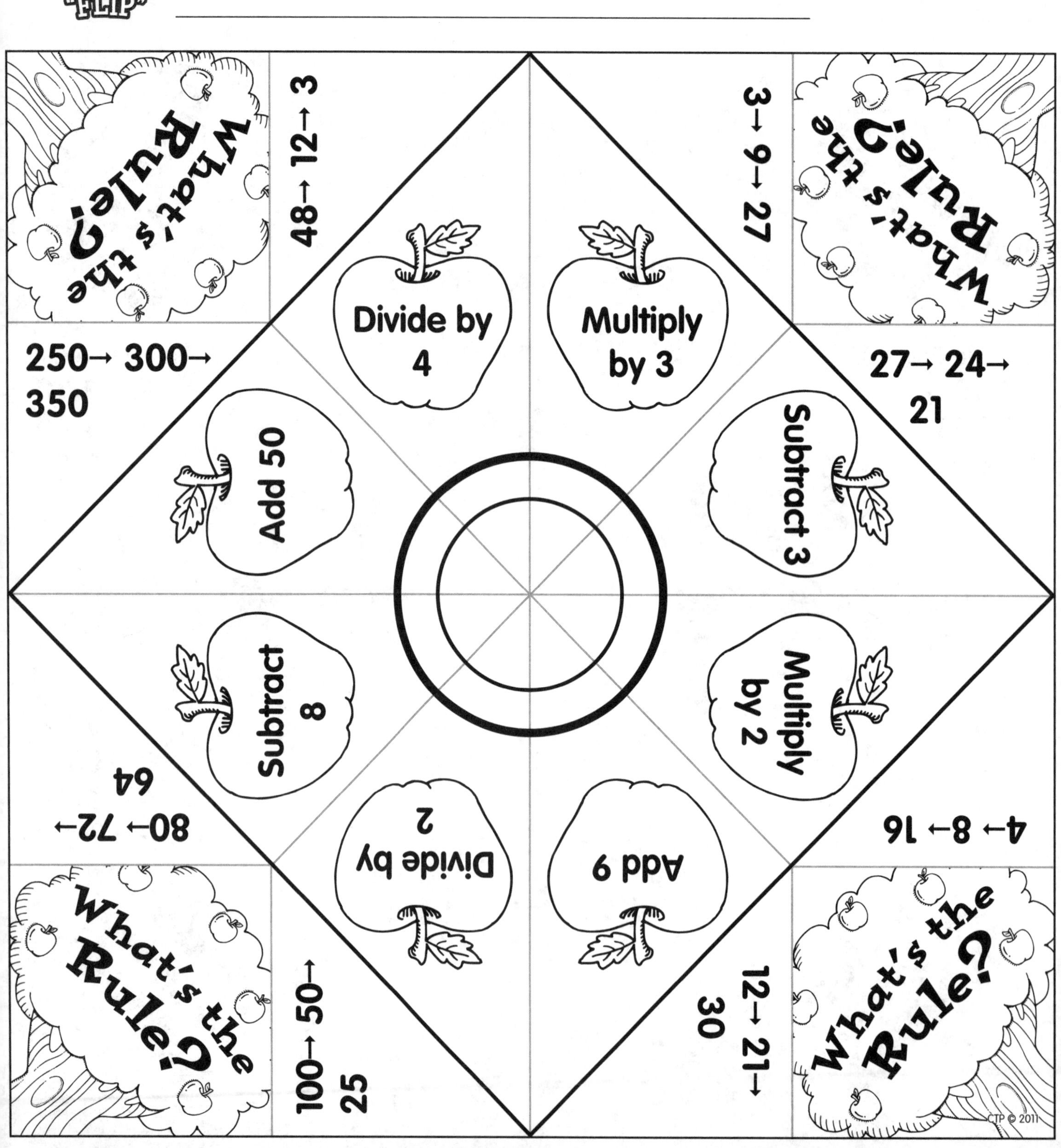

Name ______________________ Date ______________________

Order of Operations

Hint: 1. Solve the operations in parentheses.
2. Multiply or divide from left to right.
3. Add or subtract from left to right.

On the back of this paper, write the steps you need to take to solve the following:
9 + (48 ÷ 8) x 5 =

Order of Operations

24 ÷ (3 x 2) =

4

24 ÷ 6 = **4**

(12 – 5) x 5 =

35

7 x 5 = **35**

Order of Operations

(15 – 7) x 9 =

72

8 x 9 = **72**

14 ÷ (3 + 4) =

2

14 ÷ 7 = **2**

5 x (10 – 4) =

30

5 x 6 = **30**

2 x (8 + 2) =

20

2 x 10 = **20**

Order of Operations

15 ÷ (5 x 3) =

1

15 ÷ 15 = **1**

(144 ÷ 12) + 12 =

24

12 + 12 = **24**

Order of Operations

16 Name ______________________________ Date ______________________

Find the Variable

Hint: To solve $2y = 12$, you can:

Use a related multiplication fact. Think: What number times 2 equals 12? y must be 6.	OR	Use a related division fact. $2 \times y = 12$ $y = 12 \div 2$ $y = 6$

After you "FLIP"

If $y = 3$, does $11y = 34$? ____________

Explain. ____

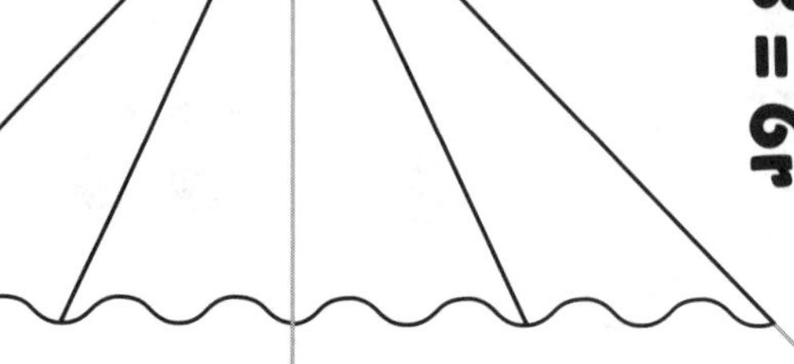

18 = 6r

10q = 50

56 ÷ 8 = s

q = 5

s = 7

y = 11

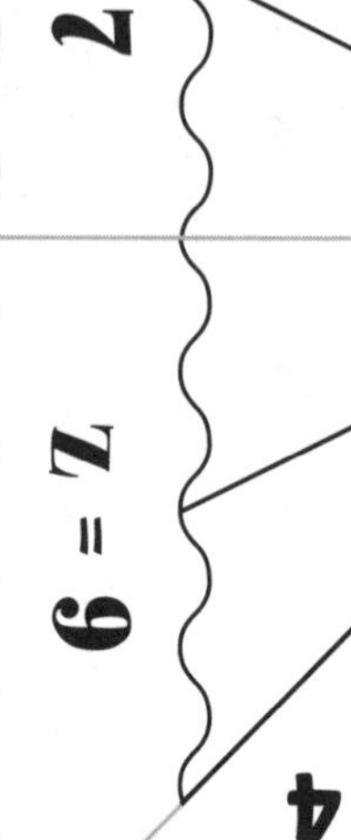

88 = 8y

36 ÷ z = 4

t = 5

g = 7

25 = 5t

21 = 3g

Name ______________________ Date ______________________

What's the Algebraic Expression?

Before you "FLIP"

Hint: Identify the operation. Then give the expression using the variable *n*.

2 more girls than boys
If n stands for the number of boys, n + 2 expresses the number of girls.

After you "FLIP"

Write an expression for the following: Lee is 1 year older than twice the age of Greg.

____

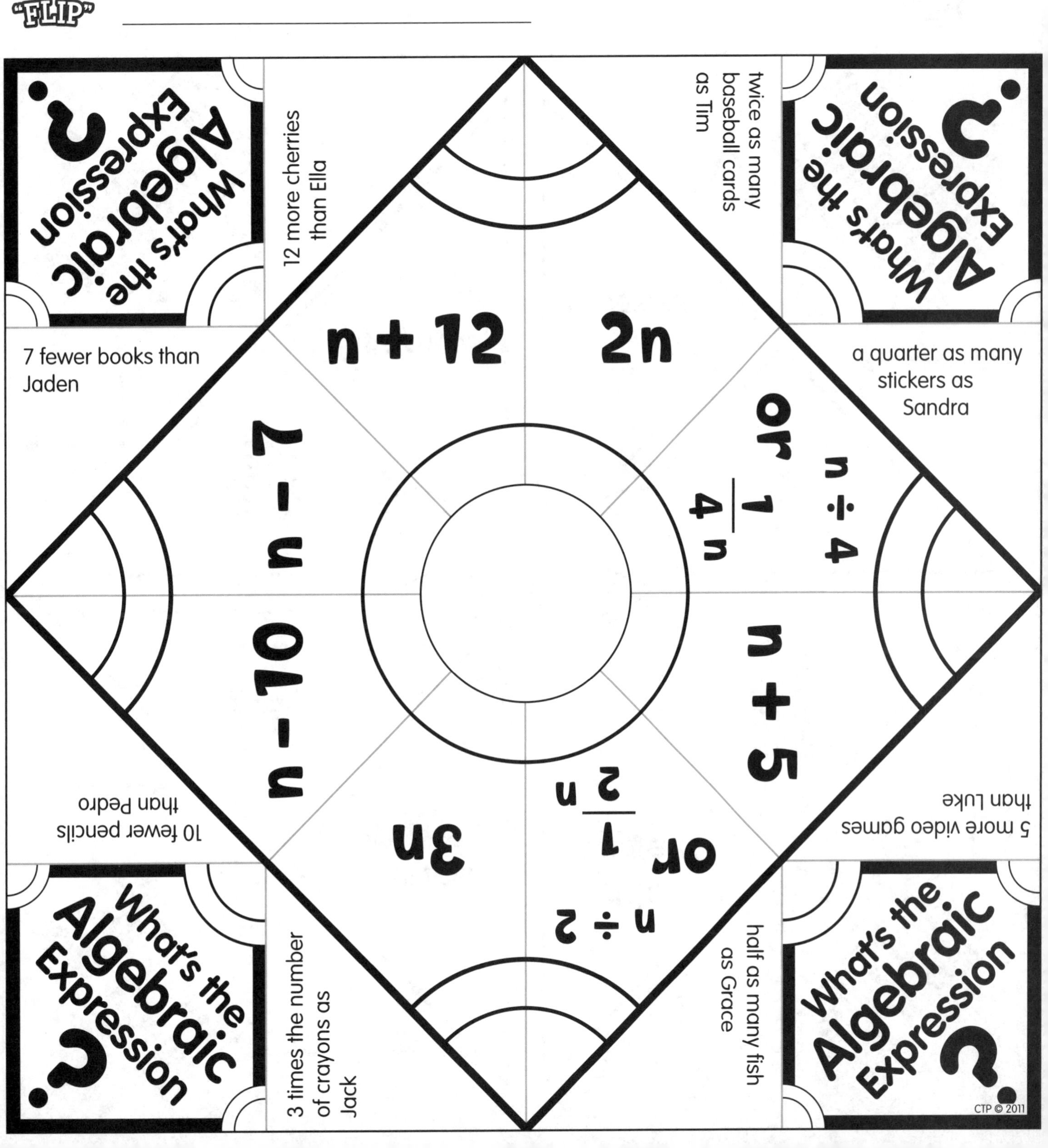

Name ______________________ Date ______________

What's the Geometric Term?

Hint: Lines can be classified as parallel, intersecting, and/or perpendicular. Angles can be classified as acute, right, or obtuse.

On the back of this paper, draw two intersecting lines that form four right angles.

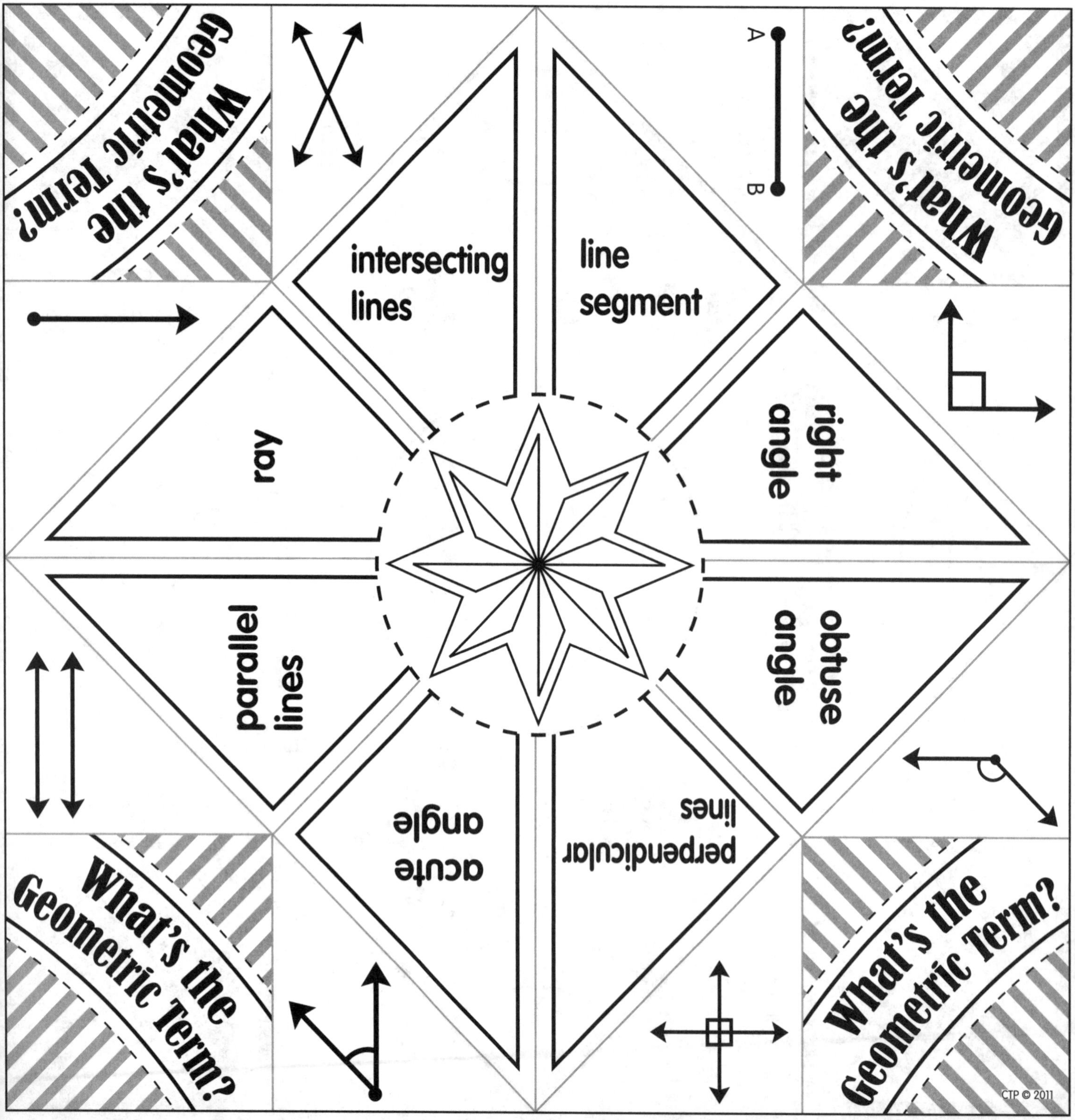

What's the Geometric Term?

A

B

intersecting lines

line segment

ray

right angle

parallel lines

obtuse angle

acute angle

perpendicular lines

What's the Geometric Term?

What's the Geometric Term?

What's the Geometric Term?

Name ______________________ Date ______________________

Are the Figures Congruent?

Hint: Congruent figures have the same size and shape.

Draw a line through the square to make two congruent shapes.

Are the Figures Congruent?

Are the Figures Congruent?

no

yes

no

no

yes

no

yes

yes

Are the Figures Congruent?

Are the Figures Congruent?

Name ______________________________ Date ____________________

Faces, Edges, and Vertices

Hint: Faces are flat surfaces. Edges are line segments where two faces meet. Vertices are "points," or corners, where two or more faces and edges meet.

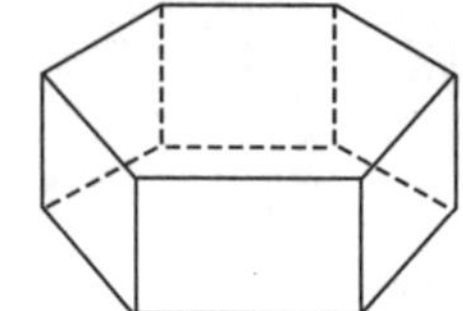

Write the number of faces, edges, and vertices for the figure on the left.

_____ faces _____ edges _____ vertices

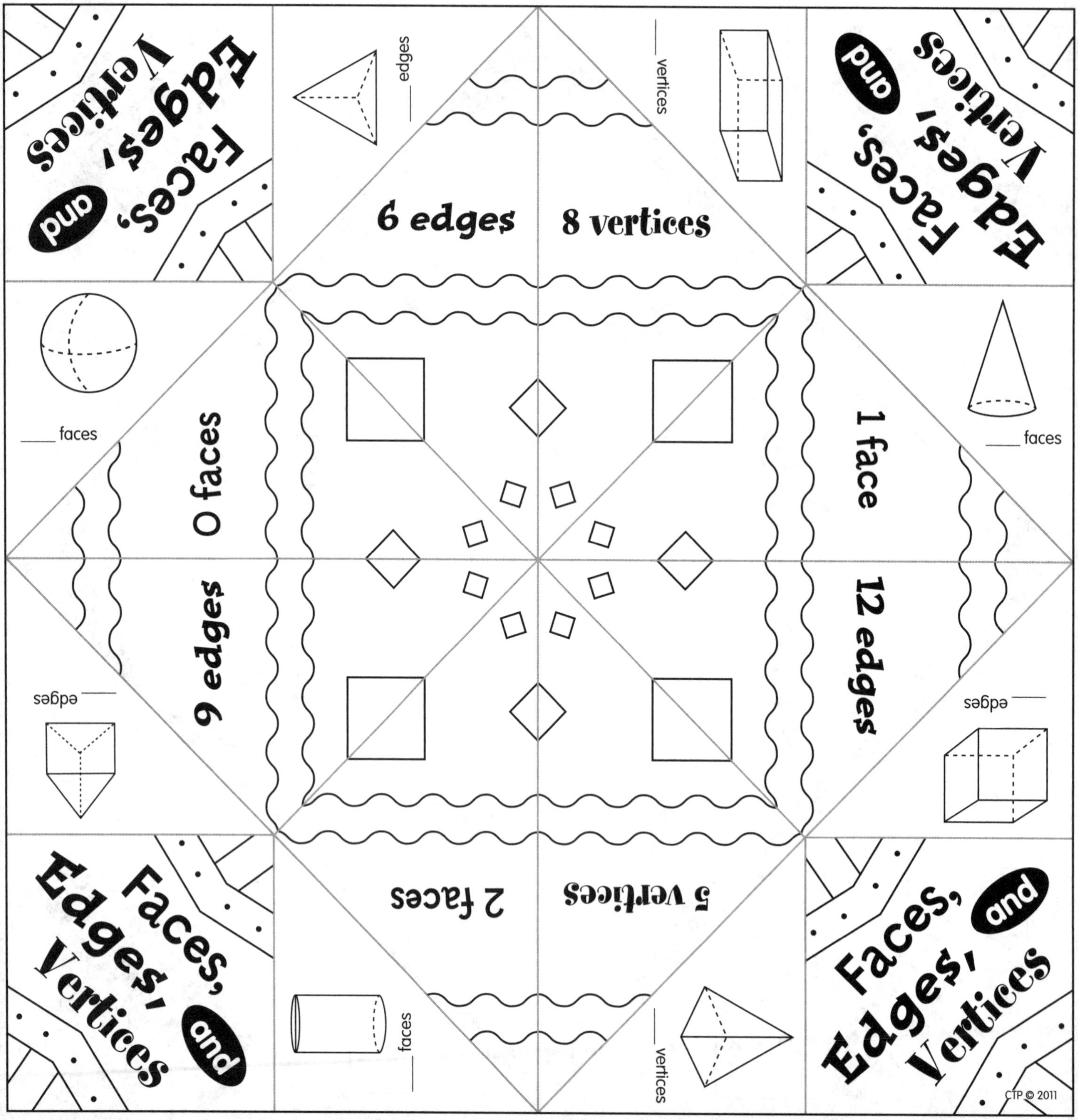

Faces, Edges, and Vertices

___ edges

___ vertices

6 edges

8 vertices

Faces, Edges, and Vertices

___ faces

0 faces

1 face

___ faces

___ edges

9 edges

12 edges

___ edges

Faces, Edges, and Vertices

2 faces

5 vertices

___ faces

___ vertices

Faces, Edges, and Vertices

Name ______________________________ Date ______________________________

Elapsed Time

Before you "FLIP"

Hint: To find the elapsed time for 5:15 p.m. to 8:55 p.m., count the hours and then the minutes.
6:15 p.m., 7:15 p.m., 8:15 p.m. = 3 hours
8:15 p.m. to 8:55 p.m. = 40 minutes
3 hours and 40 minutes have elapsed

After you "FLIP"

Calculate how much time elapses from the time you go to bed to the time you wake up in the morning.
Bedtime: __________ Wake-up time: __________ Elapsed time: __________

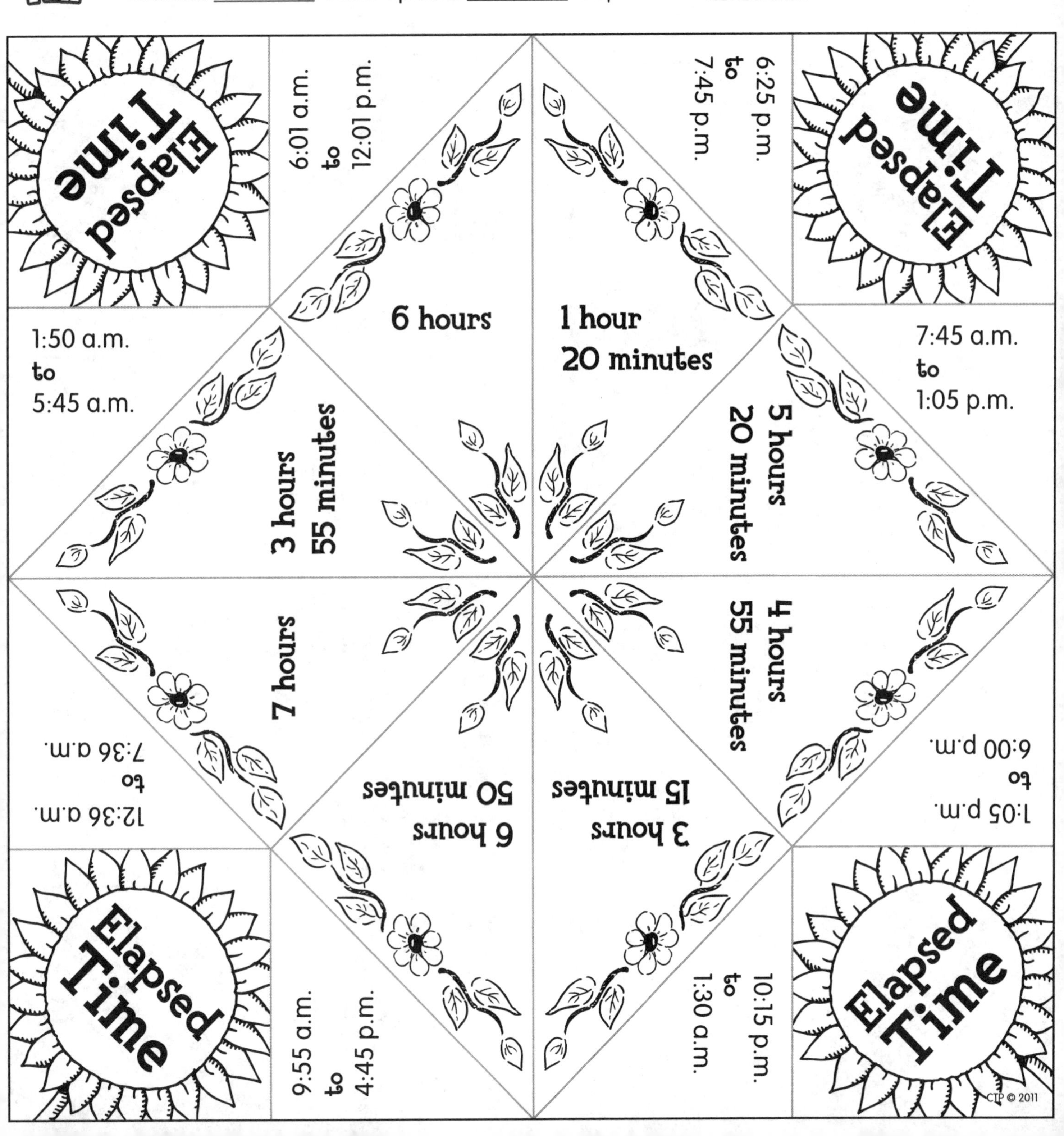

Name ______________________ Date ______________

Make Change

Hint: To make change, start with the cost of the item. Count up with the coins and then with bills until you reach the amount paid. Then add the value of the coins and bills.

You bought a game for $15.73. You paid with a twenty-dollar bill. List the coins and bills you would receive in change.

____

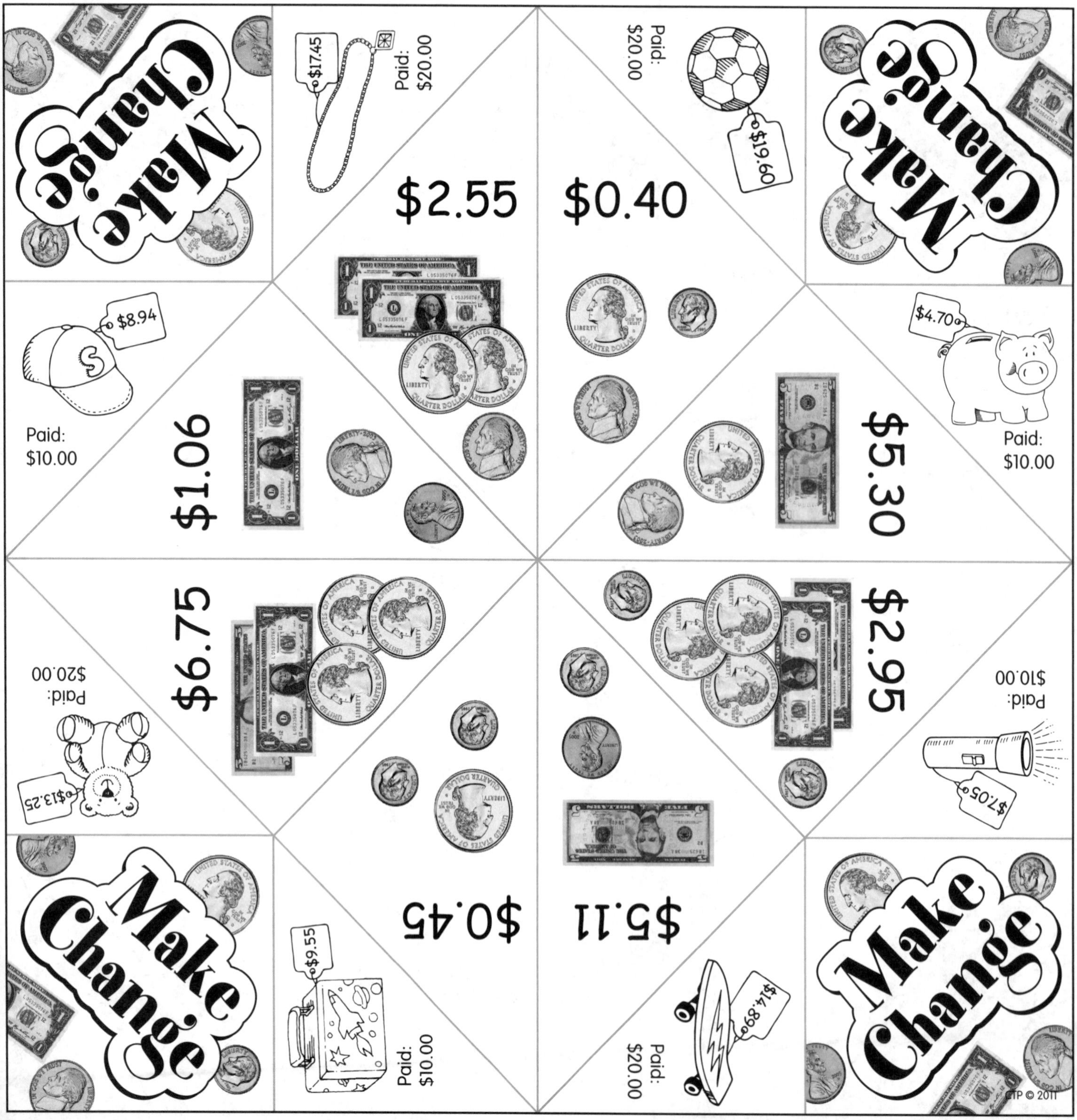

Make Change

$17.45 Paid: $20.00 — $2.55

$19.60 Paid: $20.00 — $0.40

$8.94 Paid: $10.00 — $1.06

$4.70 Paid: $10.00 — $5.30

$13.25 Paid: $20.00 — $6.75

$7.05 Paid: $10.00 — $2.95

$9.55 Paid: $10.00 — $0.45

$14.89 Paid: $20.00 — $5.11

Make Change

Name ______________________ Date ______________________

What's the Perimeter and Area?

Hint: The perimeter of a figure is the sum of all its sides. The area of a rectangle or square is the product of the length and the width.

After you "FLIP"

On the back of this paper, draw two different ways to show an area of 24 square units.

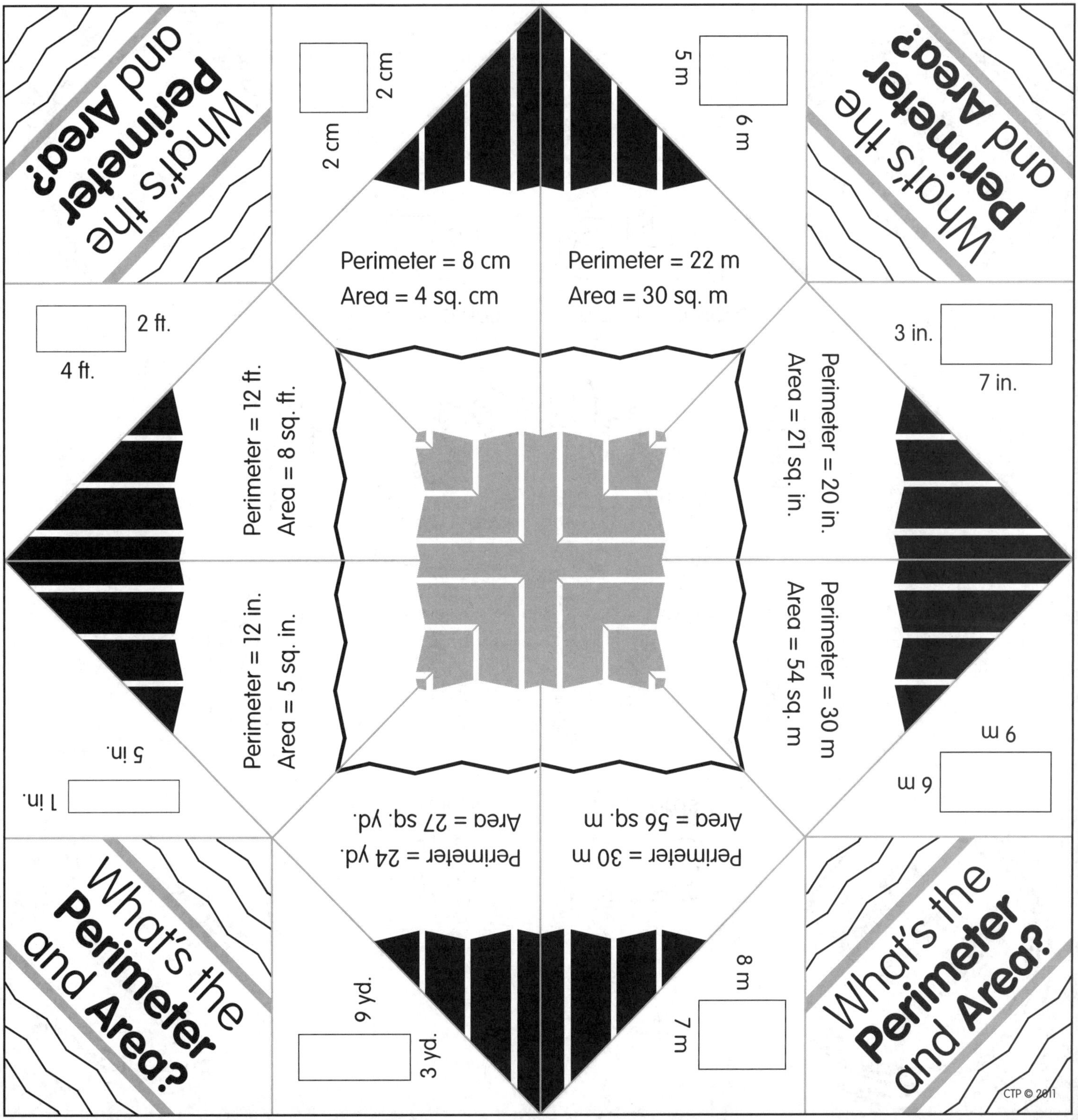

What's the **Perimeter** and **Area?**

2 cm
2 cm

Perimeter = 8 cm
Area = 4 sq. cm

5 m
6 m

Perimeter = 22 m
Area = 30 sq. m

What's the **Perimeter** and **Area?**

2 ft.
4 ft.

Perimeter = 12 ft.
Area = 8 sq. ft.

3 in.
7 in.

Perimeter = 20 in.
Area = 21 sq. in.

5 in.
1 in.

Perimeter = 12 in.
Area = 5 sq. in.

9 m
6 m

Perimeter = 30 m
Area = 54 sq. m

Perimeter = 24 yd.
Area = 27 sq. yd.

9 yd.
3 yd.

Perimeter = 30 m
Area = 56 sq. m

8 m
7 m

What's the **Perimeter** and **Area?**

What's the **Perimeter** and **Area?**

Name ______________________ Date ______________________

Find the Volume

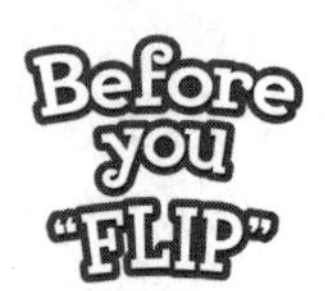

Before you "FLIP" The volume of a rectangular prism or cube = length x width x height

height
2 cm
5 cm
length
4 cm
width

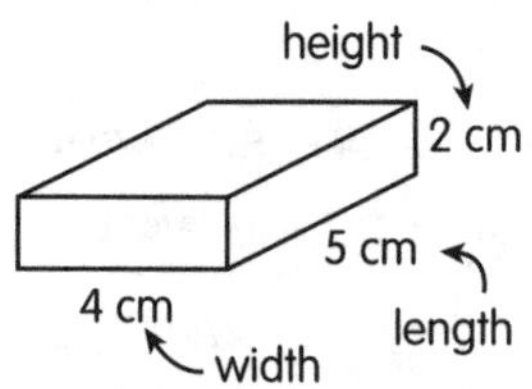

V = 5 cm x 4 cm x 2 cm
V = 40 cubic centimeters

After you "FLIP" Find the volume.

5 yd
12 yd.
9 yd.

____

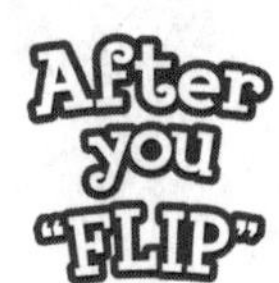

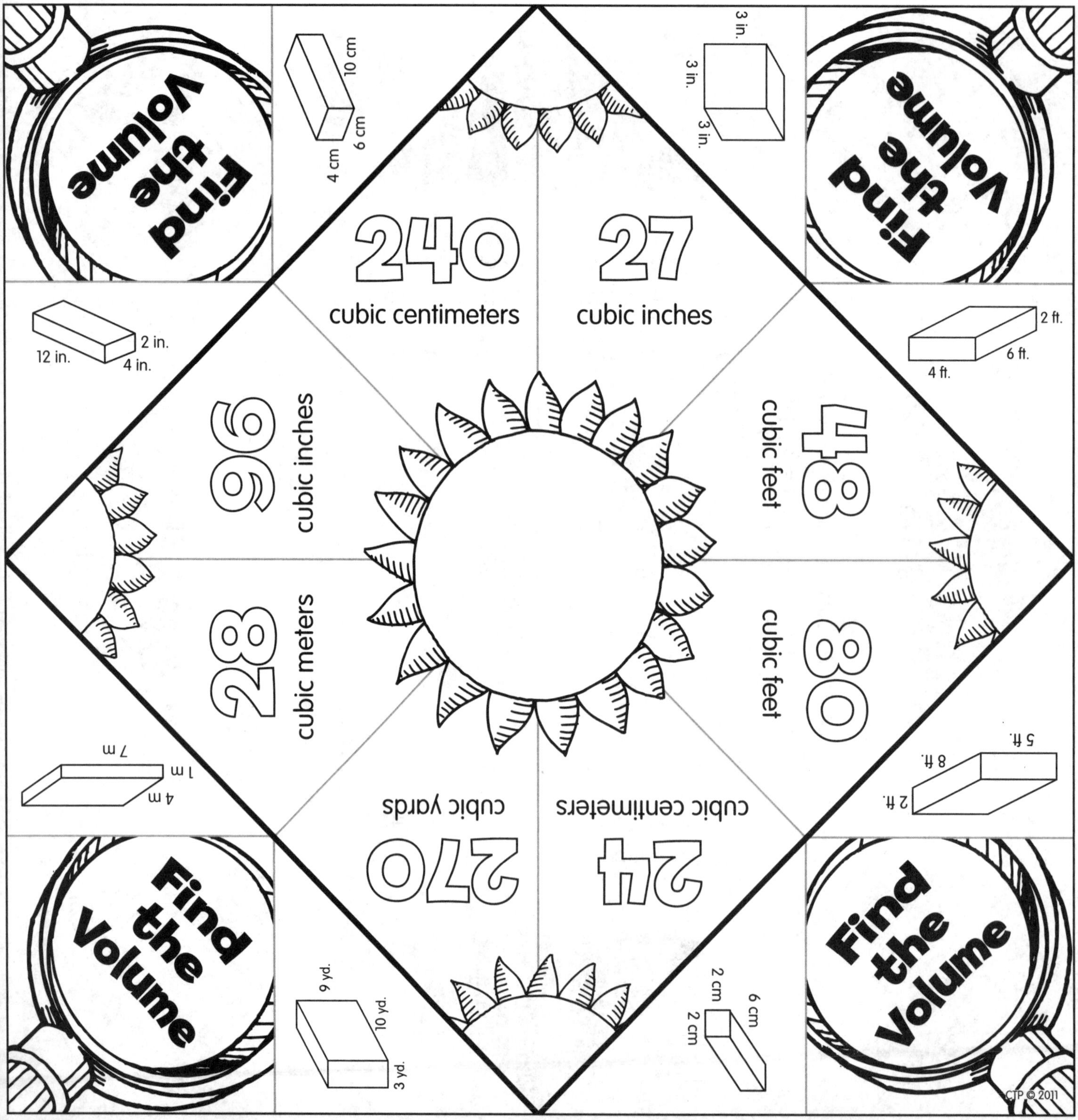

Find the Volume
Find the Volume
Find the Volume
Find the Volume
10 cm
6 cm
4 cm
3 in.
3 in.
3 in.
12 in.
2 in.
4 in.
2 ft.
6 ft.
4 ft.
7 m
1 m
4 m
5 ft.
8 ft.
2 ft.
9 yd.
10 yd.
3 yd.
2 cm
6 cm
2 cm
240
cubic centimeters
27
cubic inches
96
cubic inches
48
cubic feet
28
cubic meters
80
cubic feet
270
cubic yards
24
cubic centimeters

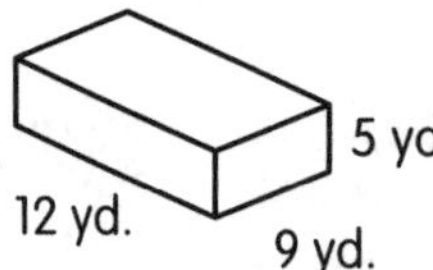